Nelton Viana

Proposal for a Computerized Document Legitimation Model

Nelton Viana

Proposal for a Computerized Document Legitimation Model

Model for Connecting Different Databases

ScienciaScripts

Imprint

Any brand names and product names mentioned in this book are subject to trademark, brand or patent protection and are trademarks or registered trademarks of their respective holders. The use of brand names, product names, common names, trade names, product descriptions etc. even without a particular marking in this work is in no way to be construed to mean that such names may be regarded as unrestricted in respect of trademark and brand protection legislation and could thus be used by anyone.

Cover image: www.ingimage.com

This book is a translation from the original published under ISBN 978-620-6-76231-7.

Publisher:
Sciencia Scripts
is a trademark of
Dodo Books Indian Ocean Ltd. and OmniScriptum S.R.L publishing group

120 High Road, East Finchley, London, N2 9ED, United Kingdom
Str. Armeneasca 28/1, office 1, Chisinau MD-2012, Republic of Moldova, Europe
Printed at: see last page
ISBN: 978-620-8-25572-5

Copyright © Nelton Viana
Copyright © 2024 Dodo Books Indian Ocean Ltd. and OmniScriptum S.R.L publishing group

Dedication

This work is dedicated to my classmates and my family in
particular.

ACKNOWLEDGMENTS

First of all, I would like to thank God for the gift of life, for my health, strength and willingness to do this course. I would like to thank UEM, which offered a pleasant and interactive environment throughout my training, the professors who taught and shared their knowledge with me, in particular my supervisor Prof. Dr. Emílio Mosse, who never lost faith in my research, guiding me with great patience and care until the end of the work. I would like to thank my family and friends who supported me and understood my absence from some family gatherings due to my studies, especially my mother Alzira Maria Viana and my wife Florinda Chivale, who supported me during the course and in this work, and my sons Edson and Mélvin, who encouraged me to keep going.

Epigraph

All education, at the moment, does not seem to be a cause for joy, but for sorrow. Afterwards, however, it produces in those who have been thus exercised a fruit of peace and righteousness (**Hebrews 12:11**).

SUMMARY

The exponential development of information and communication technologies (ICT) is seen, on the one hand, as a driving force behind the development of nations, since less effort is made to achieve great things, and, on the other hand, it is seen as a gateway to the promotion of fraud, including the falsification of documents or proof of identity, falsification of identities on social networks and other fraudulent activities. In this second approach, there has been a significant advance in the number of tools that can alter various documents in a matter of minutes, including their rapid expansion on social networks and *blogs*. In this regard, the March 31, 2022 publication on the ***EaseUS** website* clearly shows that perfection in converting PDF documents into editable formats is increasingly visible and comprehensive, highlighting the ***TOP 10*** conversion tools for Windows, Mac OS and *Online* (Zhou, 2022). Thus, a document, a job vacancy or tenders from a certain company can be easily altered and published with company logos, causing individuals to adhere to this false information and subsequently spend resources to apply. In this context, the aim of this work is to propose a computerized model that validates the legitimacy of documents and entities by presenting the reference of the document or entity in various databases connected to the model.

Keywords: falsification. validation. databases. data sharing. legitimacy of information.

Table of contents

CHAPTER 1

1 INTRODUCTION

This chapter presents the problem in the form of a contextualization, the objectives of the work, the guiding research questions and a summary of the expected results after implementing the model proposed by the author.

1.1 Background

These days, the use of ICT has been the most ideal way for organizations to streamline their daily processes. This idealistic growth trend in the use of ICT can be seen all over the world and in almost every sector. The country has been following these trends to the extent that, through modernization and computerization projects, it has implemented customer service and support services, such as the **citizen's portal**[1] **, SISSMO**[2] , etc. Noting the significant advance in the country's need to use ICT to boost the development of the public and private sectors, there is, however, also the growing wave of dissemination of documents and announcements through social networks, which often turn out to be false information, but with the unquestionable perfection of such documents.

The National Directorate of Civil Identification (DNIC), through the Mozambican government portal, announced that by August 2017, 177 cases associated with document forgery had been identified, highlighting the case of identity cards for various purposes.

produced were not from the board, but with unquestionable similarities (Moçambique, 2017).

---

[1] A platform that provides important information for citizens' lives, from basic services to job opportunities (PortalDoCidadao, 2019).

[2] It is a Social Security Information System in Mozambique, which aims to increase the speed of processing, reduce the circulation of paperwork, simplify the relationship with taxpayers and beneficiaries, etc (INSS, 2012).

The *website* of the newspaper **O PAIS** published in December 2020 (Borges, 2020), the neutralization of suspects in the forgery of documents such as IDs and car registration books, with the DNIC also unaware of their origin, which shows that there is a network dedicated to these practices.

The Order of Engineers of Mozambique (OEM) revealed on the CARTA website that there were around 60 fake engineers who were working illegally until April 2022 (CARTA, 2022).

A note published on the DW website refers to the fact that one of CEDSIF's technicians[3] took advantage of the fragility of the system that he himself was involved in designing, to issue false payment orders to the bank accounts of companies that were set up to serve CEDSIF (DW, 2017).

From the above statements, it is clear that although the trend towards the use of ICT is growing, great caution must be exercised with regard to the documents that are shared through them, and it is becoming a widespread problem, as the levels of falsification are not limited to documents, but also to entities, taking the OEM's note as an example.

Therefore, this work aims to propose a computerized model that connects to various databases in order to verify the legitimacy of documents and entities in real time.

1.2. Objectives

1.2.1. General

Propose a computerized model that validates the legitimacy of documents and entities by presenting the reference.

1.2.2. Specifics

- Identify the traditional forms of document and entity validation used

---

[3] Center for the Development of Financial Information Systems

by individual citizens;

- Identify the mechanisms used by organizations to validate documents and entities;

- Evaluate the effectiveness and efficiency of traditional ways of validating entities or documents;

- Propose the computerized model for validating documents and entities e;

- Analyze the traditional and proposed models.

1.3. Problem statement

Looking at the wave of crimes associated with the falsification of documents and entities that is recorded on television channels and in the publications of various *websites,* it can be seen that the great challenge for organizations is to adopt mechanisms to validate the authenticity of documents presented by customers and partners.

In March 2022, the company Centavo Software, SA, where the author is an employee, participated in a contest received in its institutional *email* from the company BENGA COAL MINE, but which was actually a fake contest. For the type of equipment that was required, it was necessary to make enquiries of South African providers, thus expending resources (Internet, phone calls abroad and time). This process ended 5 days after the notification was received, with the face-to-face validation of all the false documents that were shared by *email* with BENGA's offices.

The above experience, and the fact that the author of this article was part of the process of validating such a tender, has made it increasingly clear that care in relation to documents that circulate must be part of the daily practices of professionals in different areas and, on the other hand, it has become clear that the lack of a mechanism for validating documents that circulate constitutes a huge risk for society in general.

The National Association of Municipalities of Mozambique (ANAMM) made it clear at the meeting on March 10, 2023 (09:30 - Av. 24 de Julho - Nr. 2341) that there is a huge weakness in the revenue sector, particularly for the IAV which, by its nature, can be paid in any municipality and valid in others.

However, when the municipal police officer comes across a situation where a payment has been made outside his municipality, he simply ignores it, in other words, he pretends that the IAV is valid, thus opening the way for the forgery of this document which, by its nature, its fine rates (forgery) can be valued at 40,000.00 MZN (Forty thousand meticais).

Looking at this specific problem, ANAMM says it is the sector's fault for not being able to validate the taxes of other municipalities, but it also assumes the loss of revenue from all the vehicles that circulate using a supposedly false tax.

Given these scenarios, it is clear that there is a need to find a mechanism for validating documents in an agile way, as is the case with municipal police officers who, due to the nature of their work, the driver's time is precious, and the possibility of resorting to telephone calls to validate these taxes is inefficient.

1.4. Research questions

☐ What mechanisms do citizens and organizations use to verify the authenticity of documents and entities?

☐ What changes will be made by adopting the computerized model for legitimizing documents and entities?

1.5. Expected results

1.5.1. Importance of the work

With the implementation of the model proposed by the author, it is hoped to validate in real time all the documents and entities in the databases mapped in the model, validating their origin and, in general, considerably reducing the circulation and use of supposedly false information, not to mention that society itself will be educated to validate the information or document in its possession whenever necessary, thus discouraging the mentors of these practices.

1.5.2. Work motivation

Currently, there has been a considerable increase in cases of falsification of documents and entities. Through television channels or news pages, we see day after day cases related to the falsification of documents and fake agents who impersonate employees of organizations in order to achieve their goals. These cases all involve a waste of resources, both money and time. Therefore, seeing all these cases and without even being able to help the people who are going through these situations, we came up with the initiative to create a computerized model that aims to map organizations' databases in order to validate their documents, such as badges, certificates, invoices, tenders, advertisements, job vacancies and other documents.

1.5.3 Modeling the proposal

The model created essentially aims to connect to various databases through *links* previously provided by the institutions, mapping tables and attributes involved for each type of validation of this company and the subsequent introduction of the reference by the end user for the validation of such information, see Figure: 1.

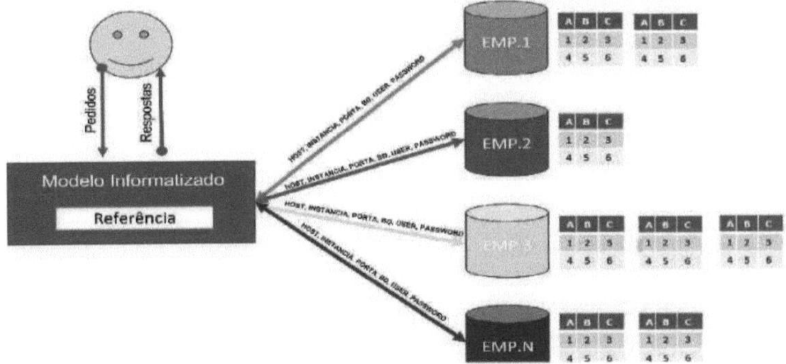

Figure 1: Implementation diagram of the proposed model (Source: Author).

You can see in the figure above that the client makes validation requests to the model, indicating the type of entity (organization) it wants to query. In turn, the model validates the data submitted to the databases of the chosen entity, according to the tables mapped when the entity was registered.

It should be noted that, in this model, the connection is made directly to the database and with only read permissions to preserve the integrity of the information consulted, not requiring any type of development or adaptation of the code for each type of validation. This makes it easier for entities to develop the module to accommodate the model's requests *(API)*.

1.6. Work organization

This work is organized into chapters as a way of situating the reader's interest in a given subject, and in this case contains six (6) chapters:

- **Chapter I, the introduction**: This chapter contextualizes the topic, presenting the problem to be solved through the objectives, the expected results of the proposal, the guiding research questions and a brief presentation of the model.

- **Chapter II, the literature review:** The second chapter provides a conceptualization of the terms most used in the work, seeking to provide evidence of the need to carry out the work, through examples of other similar initiatives and highlighting the differential of this work in a special way.

- **Chapter III, methodology:** This chapter focuses exclusively on how the research was conducted, including the modeling of the solution in question.

- **Chapter IV, the case study**: This chapter makes a general comparison between the current situation regarding how documents and entities are validated in organizations in an approximate way (an average) and the methods used by citizens.

- **Chapter V, Presentation of results and discussions:** This chapter essentially compares the author's initial idea and the result of the research carried out, both documental and that provided by the two questionnaires (for organizations and for citizens), justifying the real need to carry out the work.

- **Chapter VI, conclusions and recommendations**: This chapter is concerned with justifying the objectives set out in the work, answering whether or not the aims of the objectives have been achieved and, giving recommendations for future research related to the work.

CHAPTER 2

2. LITERATURE REVIEW

This chapter covers subjects related to the concept of documents, types of documents and their usefulness, falsification or alteration of their authenticity, the location of the work (total population and companies), the theories supporting the work and examples of work similar to this.

The choice of these concepts was guided by the nature of the work, seeking to understand the main object of the work (document) and how this document is used or shared in institutions and by citizens in their singularity.

In relation to the inhabitants and companies of the place where the work is carried out, they constitute the justification for the choice of the study samples, which will substantiate the need to carry out the work.

2.2. Document concept

According to Oliveira (2014), a document is seen as all information recorded on a material medium, which can be used for a retrieval process, such as consultation, study, proof and research. A document is reliable evidence that aims to prove facts, phenomena, ways of life and thoughts of man at a particular time or place.

According to the Infopédia *website* (Dicionários Portos Editora), a document is any object produced in order to reproduce or represent a person, a fact, a statement or an event. This publication also refers to a document as a piece of writing that serves as proof or attestation. In the computer approach, it is considered to be a file containing data generated by an application (word processor, spreadsheet, database, etc.) (Infopédia, 2022).

The Virtual State website summarizes a document as any information recorded on a medium (paper, microfilm, computer) (VIRTUAL, 2022).

Looking at the definitions of Oliveira (2014) and VIRTUAL (2022), it is clear that a document can be considered any type of record, and this type of record must necessarily be made on a medium. In this context, an engraving or writing on conventional paper are both considered documents due to the fact that they are considered records regardless of their nature.

According to the *website* **Cursos Escola Educação**, a document is any information recorded on a medium and used for study, proof and research (Educação, 2014).

Oliveira's (2014) second approach converges with the definition given in Infopédia's (2022) second approach by stating that the document serves as proof of a phenomenon, entity or fact. This convergence is supported by the publication on the **Cursos Escola Educação** *website*, which states that the document is produced and used as evidence.

In these approaches, there is a certain inclination to generate documents for the purpose of validating facts, phenomena or entities.

Due to the nature of this work, the main focus is on document approaches aimed at creating documents for the purpose of validating facts or entities.

Thus, the need to record information and then validate it brings credibility to the issuing entity, as well as the entity receiving (consuming) the information, insofar as there is a mechanism to prove that the information really is true.

2.2.1. Types of documents and their purpose

There are various types of documents according to their nature and purpose, and the classification can often depend on the purpose or scope of study, i.e. according to the publication on the GRUPO GERENCIAR website, the types of documents are only defined in 4 classifications: according to species, genre or format, subject and relevance (GERENCIAR, 2020):

- ☐ According to species, documents are grouped according to the origin of the information, i.e. how this information was produced;

- ☐ Genre or format, documents are thus classified by grouping together those in the same format (physical or electronic);

- ☐ Subject, in this classification documents of different formats can be grouped in the same classification if they have the same content or subject e;

- ☐ Finally, they are classified according to the relevance or usefulness of the information.

In classification, according to GERENCIAR (2020), what matters is the grouping of documents and not the breakdown of each one.

Oliveira (2014) looks at the types of documents as the breakdown of each document produced (letter, book, magazine, photograph, stamp, film, etc).

According to Educação (2014) the classification of documents should be looked at for both grouping and discrimination, adding the fact that the classification of official documents are those produced in institutions or organizations for administrative purposes (invoices, receipts, etc.).

Looking at the classification according to Educação (2014), it can be understood that documents can be grouped in order to discriminate them according to their subject or relevance. Thus, the relevance of the document will dictate whether it needs to be validated or not, for example, a letter written to a friend will not have the same relevance as a letter of approval to a certain entity, even though both have the same nature.

The new classification according to Educação (2014) draws attention to the fact that these documents are considered official (administrative), which means that they must be circulated by agents within these institutions, or that they are somehow used by authorized persons, in which case there is a need to authenticate their use. Therefore, looking at this need, it is clear that the existence of mechanisms for validating documents, especially official ones, must be seen as a priority action to preserve the integrity of institutions, at the risk of these documents being falsified and having catastrophic impacts on society.

2.3.   Falsifying documents

Article 322 of Section I of Mozambique's penal code considers the falsification of a document to be a:

1. Fabricating a totally or partially false document;

2. Imitating, pretending or abusing another person's handwriting, signature, firm, initial or sign;

3. Making an assumption in a deed about the intervention of persons who did not appear in it, or attributing to those who did intervene statements that they did not make, or different from those they actually made;

4. Falsifying the truth in the narration or declaration of facts that are essential for the validity of a document, or those that it is intended to certify;

5. Change the real dates;

6. To make any alteration or interspersing in a true document that

changes its meaning or value;

7. Certifying or recognizing false facts as true;

8. Drawing up a transcript, certificate, authentic copy or public form of a purported document, or one which states something different from what is in the original;

9. To include any act in a protocol, book or official register, or to register, without it having legal existence, any act of the nature of those for which the law establishes registration, or to cancel that which must subsist;

10. To make provisions, obligations or releases in any deed, title, diploma, instrument or writing, which by law must have the same faith as public deeds;

11. Adding to, changing or diminishing any part of the document, after it has been completed, in such a way as to alter its substance or intention by adding to, diminishing or changing the provisions, obligations or disobligations, or the facts which the document is intended to certify or authenticate.

According to the publication on Efomento's website, Article 297 of Law No. 9.983 of 2000 refers to document forgery as tampering with all or part of public documents or altering genuine public documents (Efomento, 2022). The same publication also raises the issue of altering dates, signatures and issuing illegal certificates from existing institutions, without having been produced by them.

It is clear from the above statements that any act (written or spoken) that involves altering part or all of a document constitutes forgery, which is also seen as imitating a genuine document, but one that was not produced by the firm or entity. Thus, since the risk that institutions run in relation to the documents they produce is evident, there is a dire need to adopt mechanisms to validate the authenticity of documents circulating in these institutions.

2.4. Entity

The Infopedia website defines an entity as that which has an independent existence or an individual who occupies a place considered important, or an organization established for social, political or economic purposes, an institution, etc (Infopedia, 2022).

From a philosophical point of view, it is defined as the essence considered in itself and not as existing in an individual being.

According to CONCEITO (2013) the entity is seen as any collective or body that can be considered a unit, and is used to refer to a corporation or company that is taken to be a legal person.

The dictionary on the SAGE website defines an entity as a public or private legal person with its own legal personality (SAGE, 2022).

On the *webpage* **of meanings**, entity has the meaning of individuality, being, that which constitutes the essence of something (SIGNIFICADOS, 2022).

The definitions above focus on the question of individuality, existence and essence. In this, it is clear that entity can be seen as something that exists and occupies only one space, substantiated by the definition of Infopedia and Significados in representing entity as something that has existence and individuality.

The term entity is used in this work to refer to the unique existence of a person or organization, in which case it can be represented by its own unique identity, i.e. the identity card number of person A cannot be the same as that of person B, and so it is with institutions.

It is not enough to validate the existence of a certain document belonging to entity A; other elements such as facial recognition or other forms must be proven to entity A. Thus, the existence of a document belonging to entity A must be used exclusively for the purposes of the same entity without, however, adulterating any of the data to belong to another entity (referred to in this paper as document falsification).

2.5. Interoperability

According to SINFIC (2006), interoperability is understood as the ability to transfer and use information uniformly and efficiently between various organizations and information systems, and can be represented at various levels: " *Hardware* and/or *software* " Network integration " Systems integration " Technology integration.

The interconnection of technologies as a way of taking advantage of interoperability is seen as the possibility of linking computer systems and services, whereby systems and devices have the power to exchange data reliably and without associated costs (Almeida, 2019).

According to the quote in Ravelli's dissertation (2003), interoperability can be seen as the ability of applications running on different computers to exchange information and operate cooperatively using that information.

From these definitions, it is clear that applications installed in different environments have the ability to exchange information for a specific purpose or objective. Therefore, the sharing of resources *(hardware and software),* or the integration of technologies, is the main focus of the term interoperability, always seeking to solve a data sharing problem. In this work, the major problem is the need to share information between different databases, in order to respond to the need to validate a document or the existence or non-existence of an entity.

Solving this problem necessarily involves creating a model that integrates different databases, trying to solve the problem of interconnecting heterogeneous databases, as cited in CIN-UFPE (2007), when referring to the need to solve the problem of interoperability between databases created in different environments and running on different computers, and the challenge is exactly to create the integration model. Therefore, this work considers the challenge of connecting these databases from different environments, using the model proposed in the work. The model must be able to connect different databases regardless of their version, nature and composition (tables and attributes).

2.6.   Theories of work support

The use of Information Systems (IS) in organizations has taken significant strides these days, due to the need to automate processes that aim to improve the working environment in these organizations and consequently increase the organization's competition with others.

For Da Cunha, Júnior and Lucian (2009), information systems and organizations are closely linked due to the fact that, on the one hand, systems provide vital information for the sectors of organizations and, on the other hand, for organizations to benefit from new technologies, they must be aware of the influences of information systems and open to them.

There is a clear trend towards the use of information systems as the perfect way for organizations and society in general to succeed. Cited in an article by Lessa, Negash and Belachew (2016), Walsham and Sahay (2005) point out that e-government has shown significant advances around the world, by helping governments to become more efficient and more effective, thus strengthening relations with citizens, companies (organizations) and other government spheres.

Although significant progress has been made in considering information systems to be ideal partners for the success of organizations, there have been several projects which, for adverse reasons, have only gone down in history as failures in the development and implementation of IS, especially in public institutions. This idea is substantiated by the authors Da Cunha, Júnior and Lucian (2009), who state that although business organizations see ISs as fundamental tools for increasing competitiveness, public organizations face great difficulty in adopting them and, consequently, in adapting to new working practices, which are geared towards resistance to change.

Lessa, Negash and Belachew (2016) again cite the author Heeks (2008) who, according to his study, around 35% of e-government implementations in developing countries can be classified as total failures, i.e. started and abandoned immediately.

In the search for reasons for failures in the implementation of IS (most visible in public institutions), Lessa, Negash and Belachew (2016) cite Bhatnagar (2000), who states that there is a lack of commitment on the part of public leaders and managers. In the same vein, the same authors cite Aichholzer (2004) who considers the problem to stem from poor management of long-term sustainability risks.

Looking at the above approaches and, in a way, requiring a change in the way of working after the implementation of the model proposed in this work, there is a need to visit the institutional and organizational change theories in order to understand how the problems listed above can be minimized, taking into account the context and structure, that is, it is necessary to understand the context in which they will be implemented (organizational context).

- the behaviors of each of the players) and structure (observing all the hierarchical levels of each institution) (Chilundo, 2020). On the other hand, the use of structuring theory is evident, in order to make the user (agent) aware of the structure.

- who in this theory is taken as a reflective agent, one who reflects on his activities for the success of the organization) that the success of the IS will only be remarkable if it meditates on its contribution to the success of the organization (Muquingue, 2020).

By applying these theories, it will follow that agents of organizations, as well as citizens, will think again before referring to the situation of falsification of any document, to the extent that they will be aware of the existence of a platform or mechanism for validating entities or documents, in use in the organization (for the case of the agent), or for the general public.

In De Araujo's (1995) approach, it is understood that solving problems related to ISs necessarily involves dividing them into smaller dimensions, i.e. complex solutions have a greater loss of energy (entropy) than smaller ones.

Taking the above approach into account, developing a model that is simple in terms of the steps taken to retrieve information can make it a real win and not become a total failure. Associated with this idea, it is clear that for the specific case of the model in reference in this work, one of the ways to make it accessible and acceptable would be to quickly disseminate it in all spheres of society, in companies as well as to citizens in general.

In short, educating society in the use of this tool by publicizing it on all news channels, designing a solution that is simple in terms of composition and handling and the awareness or commitment on the part of the heads of institutions to correctly publicize and implement the solution, taken as an Information Retrieval System, would help to reduce the risks of the model failing.

CHAPTER 3

3. METHODOLOGIES

The methodologies chapter reports on the type of research that was carried out and all the paths that led to the work, highlighting the data collection and processing techniques. Finally, all the outlines that led to the development of the model are presented, including the tools used to do so.

3.2. Research

The work essentially aims to respond to the problem of falsification of documents and entities, using methodologies based on the search for information based on questionnaires and observations. This is applied research, as it seeks to provide a practical solution to the problem in question, based on the realities experienced by institutions and citizens in general.

By not using statistical methods to analyze the results obtained, it is clear that this is a qualitative approach. It is also clear that the problem as much as the solution cannot be translated into numbers, but rather the demonstration of the model as one of the ways to solve the problem.

As far as the objective is concerned, this is exploratory research, because it relates examples of problems similar to the one we are trying to solve in the work and because it uses a case study to explore in detail the problems of sharing allegedly false documents in institutions and/or on public information consumption channels (social networks, television channels, etc.).

This work is seen as a case study in terms of technical procedures, because it provides concrete examples, looking at some Mozambican companies based in the Province and City of Maputo and also the reality of citizens in the same places. In concrete terms, the work seeks to bring concrete problems experienced by citizens with regard to the falsification of documents and entities and the same scenario experienced by companies, that is, through questionnaires to companies, we seek to know how they view the problem of document falsification and what measures are taken to minimize cases.

Thus, in order to achieve the objectives described in the first chapter of this work, data collection techniques were used, such as a literature review, observations and questionnaires.

The literature review provides the basis for conceptualizing the main objects of the research, i.e. documents and entities, including the basis for building the problem-solving model (Interoperability).

The questionnaires are divided into two groups, one for citizens, with the aim of understanding their awareness of the problem of falsification of documents and/or entities, through their experiences. The second group of questionnaires was aimed at companies, as a way of understanding the mechanisms used to validate internal and external documents.

The questionnaires were administered on the *google forms* platform, where 44 sensitivities were collected for citizens and 17 for companies, as this is a considerable number to assess the problem and the proposed solution.

The questionnaires are clear in preserving anonymity for citizens as well as companies, although some questions ask respondents to identify themselves.

In addition to the data collection techniques described above, observations were necessary for the requirements gathering phase of building the model.

The schedule below was used as a data collection and processing plan.

Table 1: Data collection and processing plan

ENTITIES	DESCRIPTION	PERIOD

COMPANIES	Sending data collection *links*	For two months
CITIZENS	Sending data collection *links*	For a month
AUTHOR	Data handling and processing	For two weeks
AUTHOR	Gathering requirements for the model proposal	For a month

3.2 Modeling and model development

To solve the problem posed in this work, the document and entity validation solution was modeled and developed using software engineering techniques (data or requirements gathering, solution modeling, development and testing).

In the data or requirements gathering phase, observations were mainly made at nearby companies such as CENTAVO SOFTWARE, SA. The observations were accompanied by immediate questionnaires, with the aim of finding out from the employees the difficulties encountered in validating external documents (from other institutions) and, in particular, trying to understand how the solution would be designed to accommodate this need.

In the modeling phase, the *Gliffy Diagrams* tool was used because it is convenient for modeling objects and accepts the introduction of external figures. Particularly for this work, more emphasis was placed on the modeling component, in order to make the solution comprehensive, as each programmer will be able to understand and code according to their convenience.

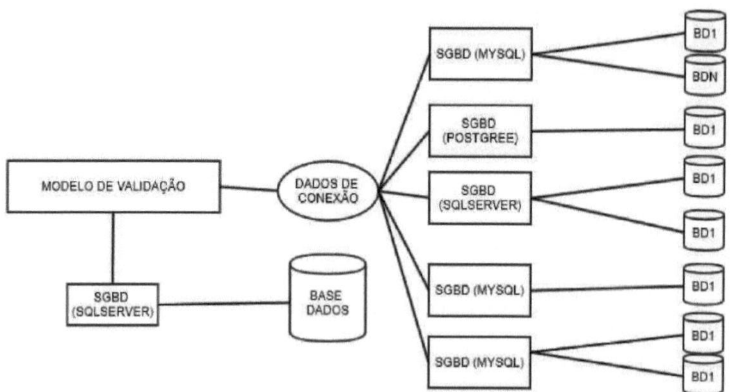

Figure 2: Implementation Model of the Validation System with different DBMS (Source: Author).

The model for validating documents or entities *(software)* was developed in the *Microsoft* programming language c-Sharp (C#). This language was chosen taking into account the author's experience, thus requiring less effort to model, design and develop the solution.

The Database Management System (DBMS) adopted by the author is *Microsoft Sql Server* version 12. In this environment, the database was created to store the communication settings for the different databases to be mapped for validation purposes. *Mysql* was also part of the connection test, in which various databases were mapped in the model and tested on different computers.

The test in two different environments *(Mysql* and *Sql Server)* was adopted as a guarantee that the model would work in databases that have conquered the market in recent years.

In general, the aim of this work is to create a standardized model for connecting different databases without the need for in-depth programming knowledge, thus making it easier for different institutions to use the model, simply by indicating the path and the database connection data: Database

Manager (DBMS), server, database, port, user and password.

1. **DBMS**: This parameter indicates the database management system used at the institution (Mysql, MS SqlServer, PostgreSql, Oracle, etc);

2. **Server**: This parameter allows you to point to a physical address of the database server (computer), via a public *IP* or domain. Note that for local connections the server will always be *localhost.* In a broader view, it is necessary to give permission for external connections from the database manager's administration page. When you have access to the physical server, you can set *limit to hosts Matching* to the percentage symbol *(%)* for Mysql users or enable remote access for databases managed on the *Internet* or in a DBMS such as *SQLServer.* As a way of guaranteeing the security of restricted access for this purpose (validation), in this specific case for remote database queries, it is necessary to map the specific access IP address, so that even for those who have access to the database and the access credentials, if the domain or IP is not mapped, they will never have access to the database.

3. Database: This parameter allows you to enter the name of the database you want to connect to;

4. **Port:** Allows you to enter the database access port on the network or server (3607 for *Mysql,* 1433 for MS *SqlServer,* etc);

5. **User:** This parameter allows you to specify the database access account *(root,* etc);

6. **Password:** Still on the database access credentials, it is necessary to indicate the password used when creating the user.

To guarantee the consistency of the data consulted by the platform, the users to be provided should only have consultation permissions and if possible on specific databases and tables.

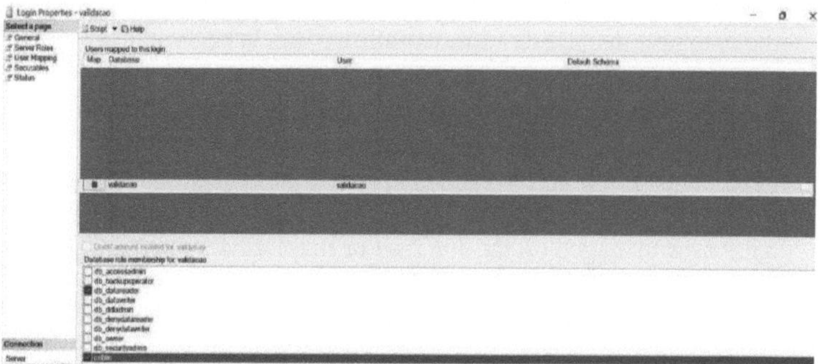

Figure 3: Example of a user with limited access to the database (Source: Author).

In addition to the connection data, the model also needs the tables used for validation and the query attributes such as the return attributes. This is why a model was created that allows connection data to be changed from any point and at any time, without the need for in-depth programming knowledge, i.e. just knowing the connection data or access to the database.

CHAPTER 4

4. CASE STUDY

This chapter justifies the choice of sample for the questionnaires, presenting the population of the place where the work was carried out, as well as the statistics of companies in the country. This chapter also highlights some examples of institutions that have platforms for validating internally produced documents and the need to harmonize them with the model proposed in this work. Specifically, bringing in examples of other platforms has made it possible to analyze the difference between the platforms under analysis and the solution proposed by the author (model).

4.1. Case studies

The work was carried out in Maputo Province and Maputo City, because it was easy to obtain data from companies and citizens.

According to the IV General Population and Housing Census of 2017, of the 27,909,798 inhabitants in the country as a whole, Maputo city had around 1,120,867, corresponding to approximately 4.0% of the country's population. Maputo province had 1,968,906, corresponding to approximately 7.1% of the country's population (INE, IV GENERAL SURVEY OF POPULATION AND HOUSING - 2017, 2019). Of this population, the work was limited to collecting the sensitivities of 50 elements, corresponding to approximately 0.002% of the total population of the two locations (Maputo province and city). Although it is a relatively small sample, the purpose of this work is to understand how citizens (respondents) legitimize documents and entities in their daily lives.

Regarding the level of education of the country's population (INE, IV GENERAL SURVEY OF POPULATION AND HOUSING - 2017, 2019), the Census reveals that 5,964 citizens had a Bachelor's degree, 163,080 had a Bachelor's degree, 15,708 had a Master's degree and 3,529 had a Doctorate, corresponding to 0.02%, 0.58%, 0.06% and 0.01% respectively, mostly comprising those aged between 20 and 80.

In terms of companies in the country, according to the study carried out between 2014 and 2015 and published in 2017, a total of 72,742 were accounted for. Of these, 15% are public companies, 70.4% private and 14.6% non-profit institutions.

According to the data from the above institutions, Maputo city had around 18,207 corresponding to 25% and Maputo province had 7,152 (9.8%) (INE, Segundo Censo das Empresas Moçambicanas, 2017).

From the data presented above in the two locations where the work was carried out (Maputo City and Maputo Province), forms for 20 companies, corresponding to approximately 0.08%, constituted the sample for the work. The representativeness of the 0.08% sample of companies in the two locations was aimed at understanding how companies validate their documents and finding the reasons for adopting the model proposed by the author.

4.2. Information retrieval tools

From the preliminary work carried out to identify ways of validating documents and entities, it was possible to identify some pages used by institutions and citizens for the appropriate purposes.

The **BAU information portal** on its website allows you to check whether or not you have a license by entering the document reference on the official website (DASP, 2020).

Figure 4: Permit validation (Source: E-BAU).

On the *website* of the Catholic University of Mozambique it is possible to check whether or not the certificate issued by the institution exists (UCM, 2021).

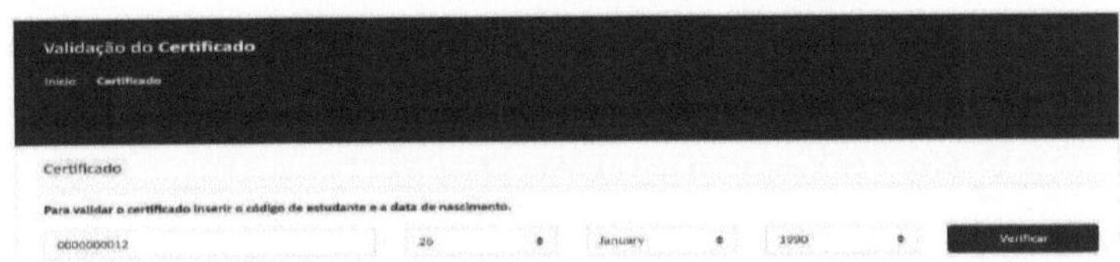

Figure 5: Certificate validation (Source: UCM Student Portal).

Still on the subject of document validation, the Tax Authority - AT, through its *website*, allows you to validate or generate the NUIT declaration (AT, 2015).

Figure 6: Validation and printing of the NUIT (Source: AT).

Discharges generated by the INSS can be validated through the SISSMO platform (SISSMO, 2011).

Figure 7: Validation of INSS Discharge (Source: SISSMO).

In short, there are so many different platforms used to validate documents. However, the citizen is obliged to be familiar with all of them, which is difficult given the immensity of the institutions and citizen services. This clearly shows that there is a tendency in the country to use ICTs to meet the needs of citizens and institutions. The proposed model essentially aims to create a mechanism for unifying existing validation platforms, by connecting the databases of each institution, through the introduction of connection data by the regulator (mediator) of each sector or area of activity.

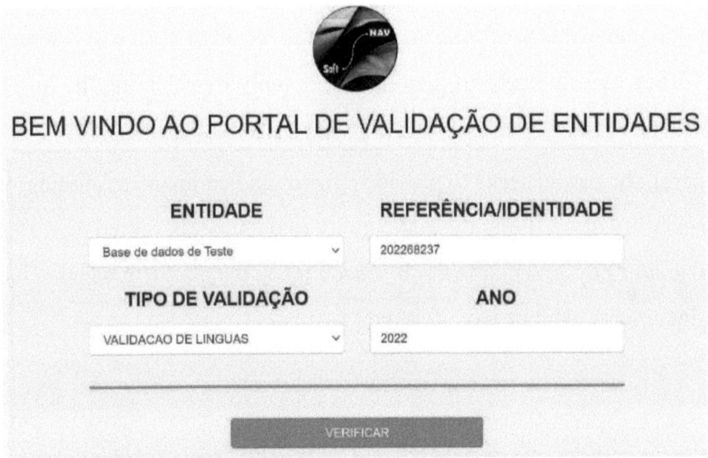

Figure 8: Document and Entity Validation Model (Source: Author).

4.3. Computerized model for legitimizing documents and entities

The model proposed by the author makes it possible to validate different entities on a single platform, simply by mapping different databases from different institutions or entities. The model was built on the basis of a (standard) *framework* that, for each new database to be integrated into the platform, no programming action is required, all that is needed is to point to the database path *(host),* the access credentials, the database name, the tables and the attributes for querying or validating the data. Unlike a connection *API*, which requires some kind of adjustment in the two applications to be connected, in this model the connection is made directly to the DBMS, requiring only the creation of access credentials for this purpose (**Figure 2**).

For the end-user *interface*, all you have to do is choose the entity you want to validate and fill in the requested data (**Figure 8**).

The advantage of this model is the possibility of concentrating all the ways of validating documents and entities, without the need for the citizen (end user) to memorize all the validation domains *(links)*. And, at some point, when they enter the platform, they can discover other forms of validation, even if they haven't heard of them before, simply by selecting the different entities (institutions) and the different types of validation.

In general, the model seeks to provide citizens and companies (entities) with the possibility of verifying the authenticity of documents shared on different digital and/or physical platforms, by checking on the platform whether or not such documents exist or have expired.

CHAPTER 5

5. PRESENTATION AND DISCUSSION OF RESULTS

N this chapter, the relationship between the theories (literature review and theories supporting the work) and the results obtained from the data collection is made. More emphasis is placed on the result of processing the data from the observations and questionnaires, in order to understand the feelings of companies and citizens about the problem of falsifying documents and entities.

5.1. Results obtained

This work is essentially based on raising awareness among companies and citizens about the need to validate documents that circulate on social networks, on information dissemination channels in public and private institutions and NGOs.

20 companies and 50 citizens were expected to fill in the forms.

In practice, only 17 companies filled in the forms, representing approximately 85% of the proposed sample. In the forms for citizens, 44 individuals filled them in, representing approximately 88% of the proposed sample.

5.1.1. Forms for companies

The form filled in by the companies had a total of 15 questions, the first 5 of which were related to identifying the company (name, type of company - public, private or NGO, area of activity, years of experience and position of the person responsible for filling in the form) and the remaining 10 related to the problem of document sharing and validation, including the possibility of adopting the model proposed in this work and the impact it has on the company and society in general.

With regard to company identification questions, around 29% of the responses came from companies in the IT sector, which means that the topic being addressed (development of a computerized model for the legitimation of documents and entities) can be better supported by the experience of these companies. In terms of years of experience in the market, around 64.7% of the companies have more than two years, which means a certain maturity in the area of document production and management, including validation itself.

Still on the questions about knowledge of companies, 13 individuals (around 76.5%) identified themselves as administrators and managers, thus showing broad knowledge of the production and validation of internal and external documents.

As far as the research questions in this paper are concerned, most of them consisted of questions related to the types of documents produced or used by the company but originating from other entities, the validation method used by the company in relation to external documents, an assessment of the problem of sharing inauthentic documents, an assessment of the model proposed by the author and its impact on the company and society in general. Specifically, only one company stated that it did not use documents such as invoices, badges, receipts, VDs, order or credit notes, diplomas and so on, representing 5.9%, and the remaining percentage for companies that responded positively.

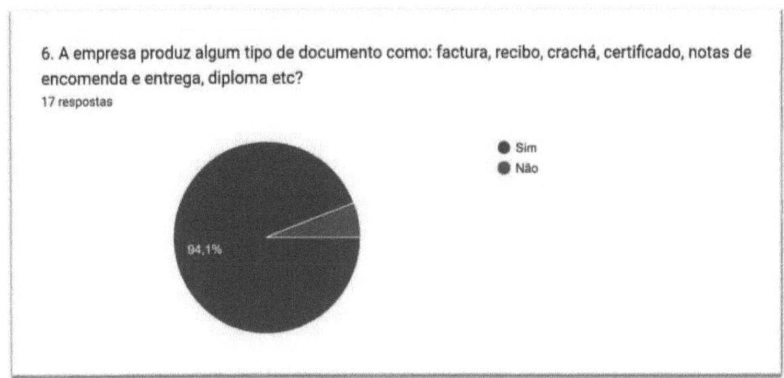

Figure 9: Answers to question 6 from companies (Source: *Google Forms).*

When asked about the need for the company to validate documents from other entities, and the way in which the company validates these documents, around 64.7% of the companies replied that they had already had the need to validate documents, with most of them having resorted to assessing the signature and stamp of the issuing institution, requesting it from the company and validating the company's letterhead.

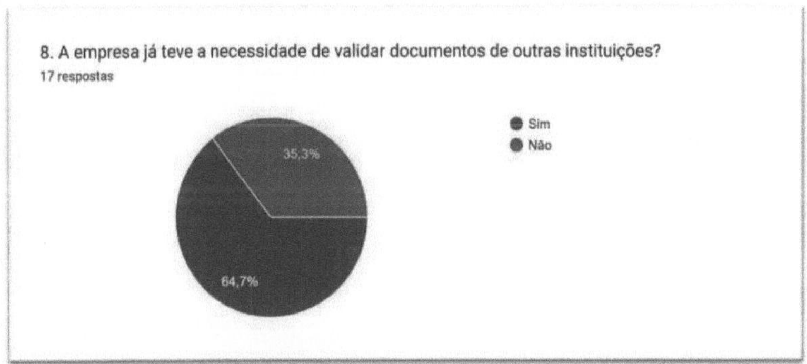

Figure 10: Answers to question 8 from companies (Source: Google Forms).

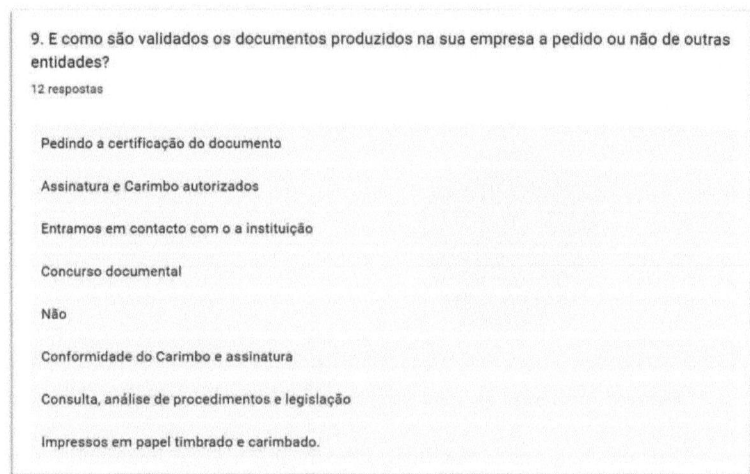

Figure 11: Answers to question 9 from companies (Source: *Google Forms).*

In the questions related to whether the sharing of false documents is really a problem and whether companies would adopt the model proposed by the author, 94.1% answered positively in the first question, with one company answering with doubt (maybe) and none answering negatively and, in relation to the adoption of the proposed *model,* 100% answered positively between the answers: *I would adopt the model, I would adopt it and recommend it to other entities and Very excellent, without comparison and,* the last option (Very excellent, without comparison) got about 64.7% of the 100%.

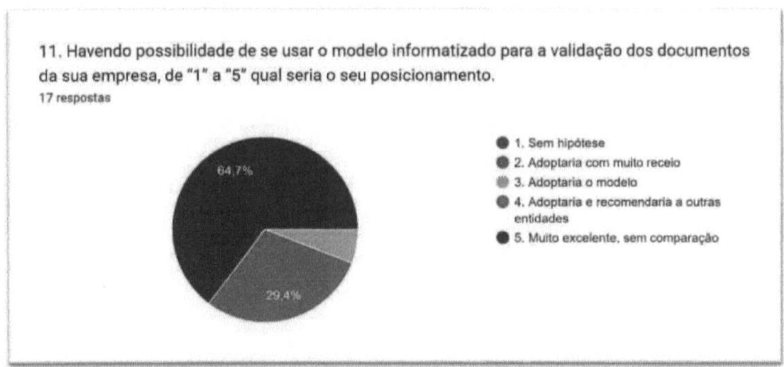

Figure 12: Answers to question 11 from companies (Source: *Google Forms*).

Regarding the companies' sensitivity to the idea that the proposed model could reduce the level of circulation of allegedly false documents, around 64.7% of the companies responded positively, 29.4% doubted this and 1 (one) company responded negatively.

Regarding the impact of the proposed model on the company, around 88.2% answered that it would have an impact on the *credibility of the information shared by the company and educated employees who are alert to fraudulent actions of a documental nature*, and around 11.8% answered with doubt *(Maybe it will have an impact, but I can't think of any), with* no company answering in the negative.

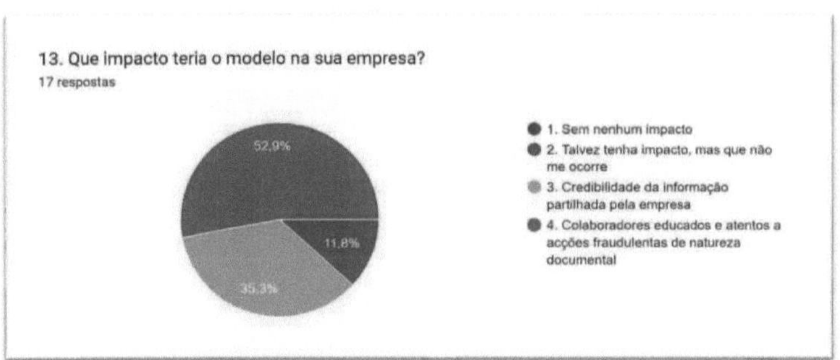

Figure 13: Answers to question 13 from companies (Source: *Google Forms*).

With regard to the impact of the proposed model on society, around 94.1%
of the companies responded positively between the credibility of the
information shared, with around 23.5% and an educated society and more
confidence in the documents circulated, with around 70.6% and only 5.9%
responded with doubt, although there were companies with negative
responses.

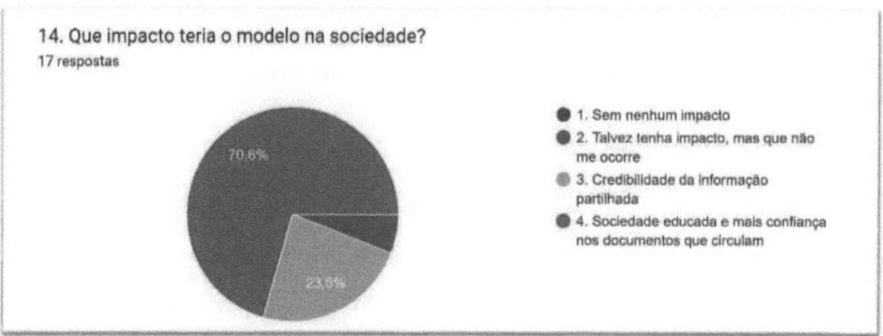

Figure 14: Answers to question 14 from companies (Source: *Google Forms).*

At the end of the questions asking for recommendations on the proposed
model, all the companies responded positively, encouraging the model to be
shared with more companies and on the need to disseminate the model, as it
is considered necessary for society.

15. Que recomendações deixaria para este tipo de iniciativas e em particular a deste trabalho
12 respostas

Recomendo que as iniciativas devem serem bem observadas de forma minuciosa de modo da evitar erros cibernéticos, devem sempre lembrar de criar mecanismo de segurança.

Que sigam em frente.
Essa iniciativa ira reduzir a circulação dos documentos falsos no mercado de emprego moçambicano e ira reduzir e automatizar o trabalho das empresa.

Insentivar e encorajar esse tipo de iniciativa nos locais de trabalho para evitar os transtornos criados pela falsificação de documentos.

Louvável

A criação de sistemas de gestão de documentação. Registo de entrada e saída de documentos, com um padrão de controlo de qualidade e rigor dos mesmos. Para que se evitem fraudes.

Recomendo

Sua brevidade

Figure 15: Answers to question 15 from companies (Source: *Google Forms*).

From the positive responses of over 50%, there is a greater need to consider the proposed model viable for educating society about the authenticity of documents and entities, as a platform that validates most shared documents will be fully available, provided that the databases of the entities in question are properly mapped.

5.1.2. Forms for citizens

The form for citizens consisted of 15 questions, the first three of which identified the individual. As the form fully preserves the anonymity of those asked to answer the questions, the identification questions only focused on age, gender and academic level, in order to understand the citizen's level of involvement in society in relation to the issue in question. Of the 44 responses to this form, only 18.2% of citizens were aged between 18 and 25, and the highest percentage was in the 31 to 45 age bracket, with around 59.1%. The remaining percentages were between the ages of 26 and 30 (20.5%) and over 46 (2.3%). It is clear from this that, as the greatest involvement came from people aged between 26 and 45, with around 79.5%, showing a certain maturity on the part of the citizens, it can be understood that the questions were answered by individuals with a certain critical vision and knowledge of the cause.

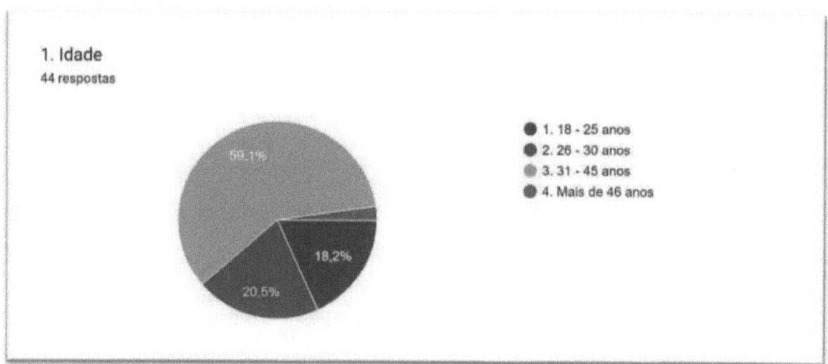

Figure 16: Answers to question 1 from citizens (Source: *Google Forms).*

The second question was merely to understand whether the greater involvement in answering the questions came from the male or female gender, although the male gender stood out with around 65.9% and the remaining percentage for the female gender.

The third question substantiates the fact that the individuals who answered the questions somehow have a certain capacity for analysis and criticism in relation to the subject, given that the highest percentage was occupied by individuals with at least a university degree (81.8%), while only 9.1% were individuals with at least a high school degree and the same percentage for a professional high school degree.

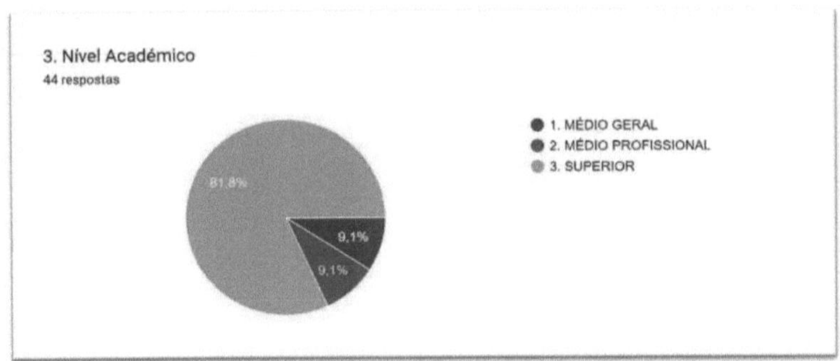

Figure 17: Answers to question 3 from citizens (Source: *Google Forms).*

In the first question related to the topic at hand (validation of information), which wanted to find out specifically if citizens knew anyone who had been in a situation where they had applied for a non-existent vacancy or had joined a fake publication, around 86.4% of the total responded positively.

On the basis of the above question, the aim was to find out whether citizens had experienced a situation where they had applied for a non-existent vacancy or a false job posting. 43.2% answered positively. In concrete terms, it can be seen that if the individual doesn't know anyone who has been in this situation, they may well have.

In question number six (6), the aim was to find out what impact the situation of a citizen applying for a non-existent vacancy or subscribing to a false publication or sharing false information had. Most of the answers showed that the impact was moral, due to sharing false information, wasting time and wasting resources (money), with these options having 37.8%, 32.4% and 16.2% respectively.

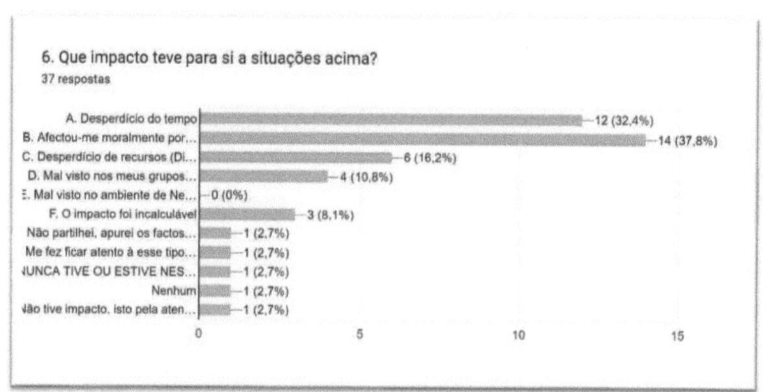

Figure 18: Answers to question 6 from citizens (Source: *Google Forms*).

Still on question number 6, seven (7) individuals abstained from answering the question and four (4) individuals answered that it didn't affect them because they hadn't shared or been in a situation of applying for a non-existent vacancy.

Question seven (7) sought to find out from citizens what mechanisms they have used to ascertain whether the information shared or adhered to is false. Five (5) individuals abstained from answering the question and two (2) answered that they had not experienced a situation where false information had been shared. The highest percentage in this question was for the answer of validating the information through the institution's publication on the falsity of the information (28.2%), 20.5% was for the answer of consulting the institution's official website and visiting the institution's premises to find out if the information was false or true. Around 12.8% were also filled in by individuals who claimed to have asked colleagues to ascertain the veracity of the facts. One (1) individual answered that they had used a credible validation platform, representing around 2.6%.

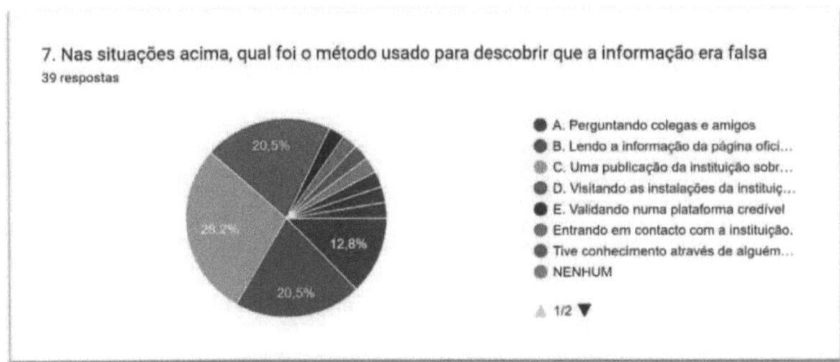

Figure 19: **Answers to question 7 from citizens** (Source: *Google Forms*).

In addition to question seven (7), question eight (8) asked how long it took citizens on average to ascertain the veracity of the facts, with only 21.1% saying that it took a maximum of 10 minutes. The highest percentage for this question was for the option of at least 24 hours to ascertain the veracity of the information and around 7.9% for 10 days or more.

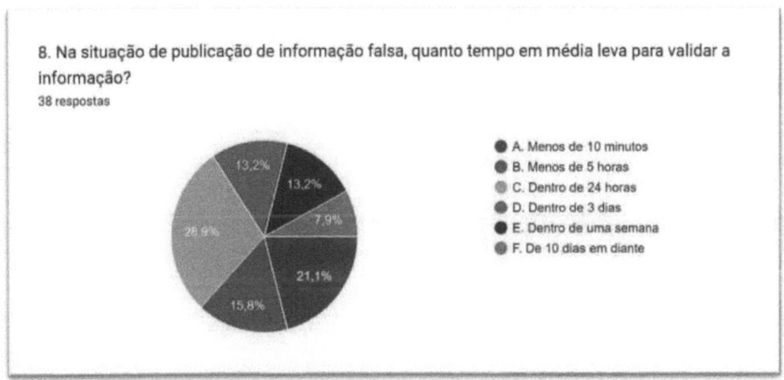

Figure 20: **Answers to question 8 from citizens** (Source: *Google Forms*).

When asked in question 9 which documents citizens would like to validate (using a platform), including certificates, driving licenses, job vacancies, competitions, identity cards, badges and invitations, around 55.8% answered that they would like to validate all of them, 4.7% answered that they didn't need to and the remaining percentages were taken up by the options[ii] *At least two* "" and[ii] *a job vacancy*"", with 37.2% and 2.3% respectively.

When asked how they validate a particular company's badge, 55.8% answered that they only look at the company's logo, 25.6% only look at the agent's clothing, 23.3% trust the agent's good intentions, 30.2% validate on a credible platform and the remaining percentages were taken up by the following options: looking at the agent's credential, nothing to do just trusting the agent, etc.

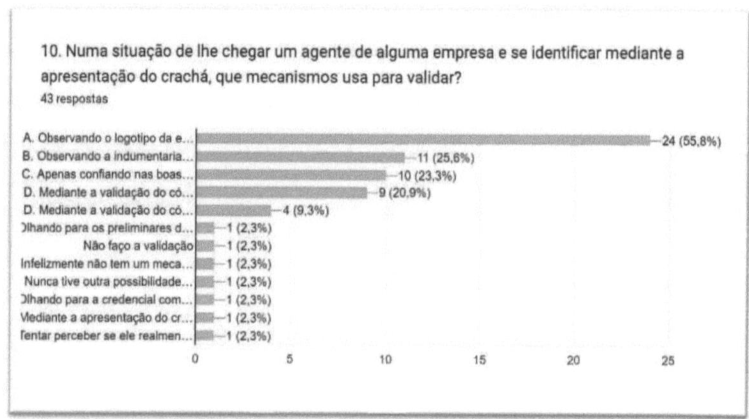

Figure 21: Answers to question 10 from citizens (Source: *Google Forms*).

Question number eleven (11) asked citizens about their level of adherence to computerized platforms for validating information, and only one (1) individual answered that there was no chance of using them. The highest percentage (around 34.1%) was for the option *"I would adopt it and recommend it to other entities"*, but 6.8% also deserved attention because they answered that they would adopt it with great trepidation.

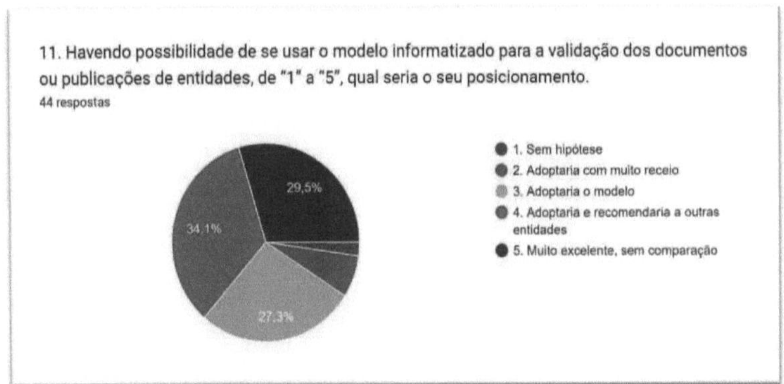

Figure 22: Answers to question 11 from citizens (Source: *Google Forms).*

Asked whether the computerized model (including the one proposed in this paper) would help to reduce the circulation of false information, one individual answered in the negative, around 34.1% answered with doubt and the highest percentage was for the *"YES"* option with around 40.9%.

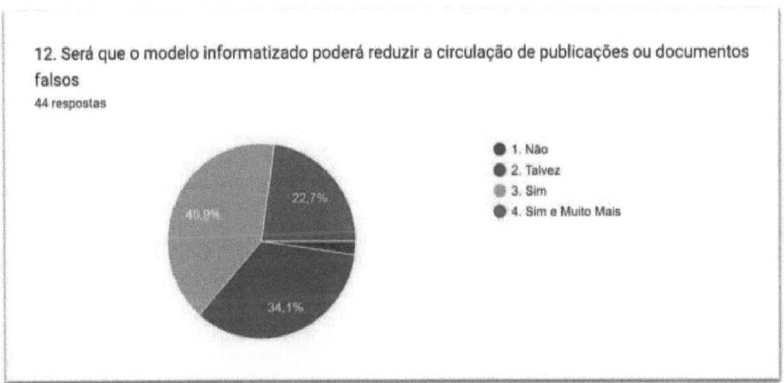

Figure 23: Answers to question 12 from citizens (Source: *Google Forms).*

On the question of the impact of the computerized model for validating documents and entities on citizens and society, all the individuals answered that it had an impact, although around 15.9% were still uncertain.

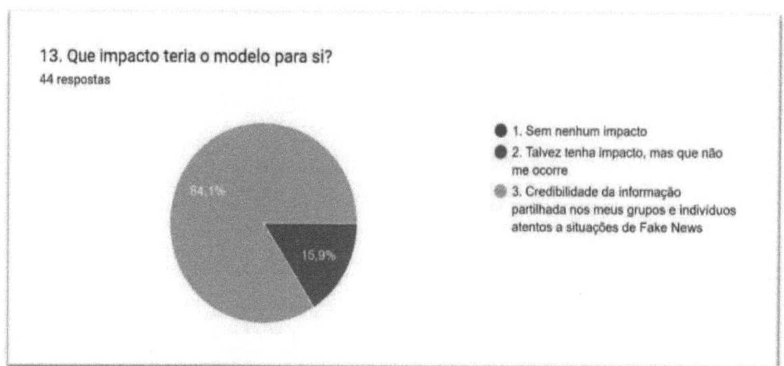

Figure 24: Answers to question 13 from citizens (Source: *Google Forms).*

With regard to the recommendations to be left on similar initiatives, the majority responded that they were social impact initiatives, recommending progress in the development and publication of the platform, thus showing the greatest adherence in the operationalization of the model proposed for validation along the lines highlighted by the author.

5.2. Discussion

The author's aim in collecting the sensitivities of citizens and companies in relation to the subject in question was to understand and essentially answer the five (5) questions below:

1. Is sharing unvalidated information a problem or not?

2. If it is considered a problem, what mechanisms are used to validate the information?

3. Would there be any possibility of adopting the proposed model to validate the information?

4. Would the proposed model reduce the circulation of false information?

E;

5. Would the model have any impact on individuals (citizens or companies) and society in general?

Looking at the 5 issues under analysis, we seek to understand whether the problem raised by the author is also seen by companies or citizens, and what solutions could be brought to minimize the problem, emphasizing the impact of the proposed model for entities and society in general.

In order to answer these questions, we used questionnaires for companies and citizens, giving each group of forms a perception of what is really considered a problem and whether or not the author's proposal minimizes it.

5.2.1. Summary of the companies' questionnaire

With regard to the first question under analysis, almost 100% of the companies answered that the lack of a validation mechanism for documents circulating in the institution or published was indeed a problem, although one company answered with doubt (maybe - 5.9%).

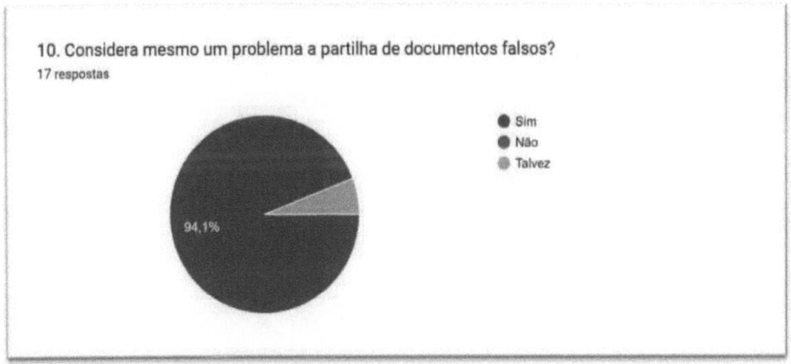

Figure 25: Summary of the first question for companies (Source: *Google Forms*).

It is clear that the concern raised by the author is also the concern of many, if not all, of the companies questioned, which in a way requires the existence

51

of a mechanism to validate this information.

As a problem, the second question sought to find out from the companies the mechanisms used to validate documents from other entities, to which almost all the companies replied that the only mechanism used to validate the information (documents) is to check the signature and stamp, which in this case shows a certain fragility in the validation of these documents, since, due to the sophistication presented in Chapter 2 - REVIEW OF THE LITERATURE, individuals in bad faith can use appropriate tools to falsify the stamp and signature of certain entities and consider them to be authentic documents in other institutions. It was also noted that information can be validated with the institutions that produce the documents (through physical visits, phone calls and emails), but in this case there is the problem of the time it takes to validate this information, which depends a lot on the distance between the institutions involved.

Therefore, with a real-time validation platform, the problems raised above could be solved in no time.

In the third question, all the companies answered that they would definitely adopt the model to validate information, including the possibility of recommending other entities.

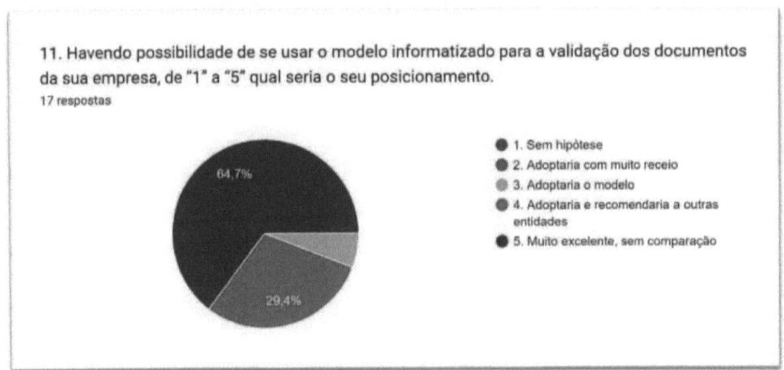

Figure 26: Summary of the third question for companies (Source: *Google Forms).*

These responses show that companies need mechanisms to validate information, as is the case with the model proposed by the author, but that there is an urgent need for it to be operationalized and disseminated.

In relation to the fourth question, only one company answered that the proposed model would not help to reduce the circulation of false information, representing only 5.9%. However, considering that around 94.1% agree with the statement, it can be considered that companies are waiting for the model to be implemented in order to reduce the circulation of supposedly false information, although around 29.4% of companies feel uncomfortable stating this categorically, as shown in the graph below.

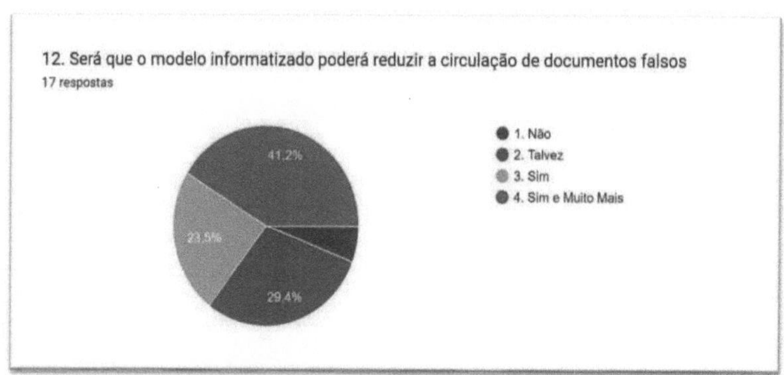

Figure 27: Summary of the fourth question for companies (Source: *Google Forms*).

In relation to the last question under analysis concerning the impact of the proposed model, around 11.8% of the companies answered that they were in doubt about the impact and the remaining percentage (88.2%) answered in the affirmative, between *"credibility of the information shared by the company"* and *"Employees educated and alert to fraudulent actions of a documentary nature",* corresponding to 35.3% and 52.9% respectively.

In this context, it can be seen that, with the implementation of the proposed model, companies would have greater reliability in the information they share and that their employees would pay more attention to issues related to validating information, thereby reducing situations of fraud or forgery of checks, receipts, invoices, certificates and other documents.

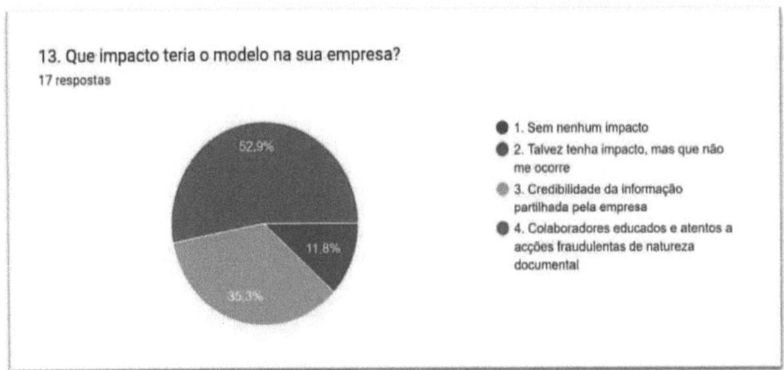

Figure 28: Summary of the fifth question for companies - Individual impact (Source: *Google Forms*).

Still on the impact of the model on society from the companies' point of view, only one company responded that it had doubts about the impact on society, representing just 5.9%. Considering that around 94.1% of the companies answered in the affirmative, it is clear that society does need similar initiatives in order to boost socio-economic development through actions aimed at reducing the level of fraud and rationalizing time.

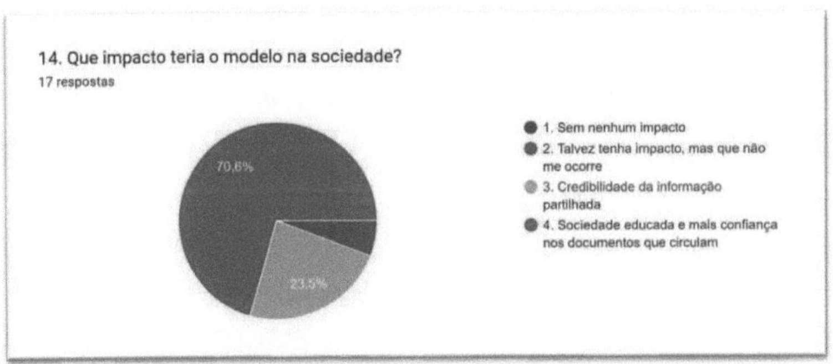

Figure 29: Summary of the fifth question for companies - Impact on society (po *nte· Google Forms)*

In general, more than 70% of the companies responded positively to the assumptions made by the author, and may consider that the sharing of unvalidated information is indeed a problem and that the author's initiative is welcome and could reduce the level of circulation of false information.

This argument is based on the recommendations left by the companies, as shown in figure 15.

5.2.2. Summary of the citizens' questionnaire

In the first question, the majority of citizens answered that the issue of validating information had cost them time, money and affected them morally, because they had shared information that they later found to be false, as shown in figure 18. Specifically, only 10.8% of citizens responded that they did not see the sharing of allegedly false information as a problem.

Thus, given the higher percentage of those who responded positively, it can be considered that a way should really be found to reduce the problem of validating information.

In the second question on the mechanisms used by citizens to validate information, only one individual answered that they use a credible platform for this purpose, representing only 2.6% of those who answered this question. This shows that there is a huge gap in the validation of information, with around 97.4% of individuals still relying on consulting colleagues, official websites of the institutions that publish the information, visits to the institutions and other options. Thus, if a validation platform existed, it is believed that it would reduce the impact caused by sharing unvalidated information, associated with wasted time and resources.

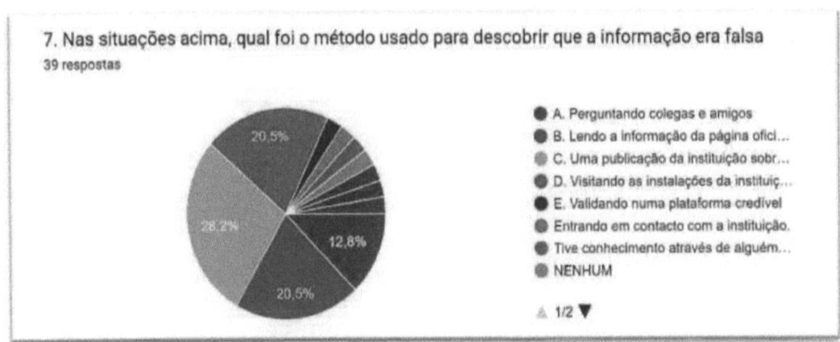

Figure 30: Summary of the first question for citizens (Source: *Google Forms*).

In the question about adopting the model proposed by the author, one individual answered that they would definitely adopt the model, representing around 2.3%, 6.8% of the individuals answered that they would adopt it with trepidation and around 90.9% of the citizens answered that they would adopt the model. Given the higher percentage of individuals who responded positively to the use of the platform proposed by the author, it is clear that society feels the lack of similar solutions to give credibility to the information shared and reduce the efforts of those who need reliable information.

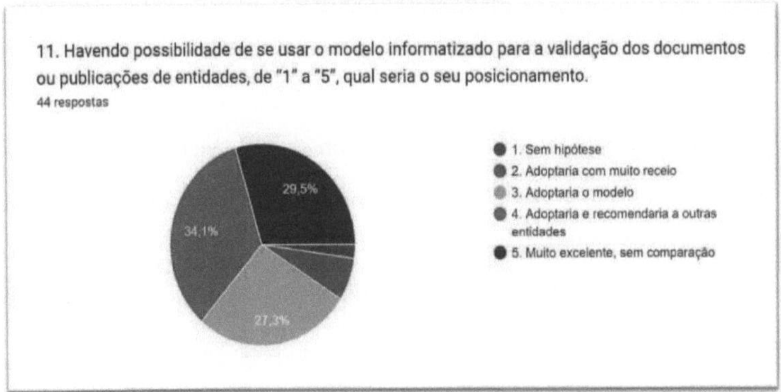

₀**Figure 31: Summary of the third question for citizens** (Source: *Google Forms*).

In the fourth question, the majority of citizens answered that the model could help reduce the sharing of false information, representing 75% of the answers, although around 34.1% of citizens answered with some doubt about the assumptions and one individual answered negatively.

However, the representativeness of the majority indicates that, to a certain extent, the existence of a computerized platform could reduce the level of circulation of false information, to the extent that information from any institution can be validated even before it is published to close friends.

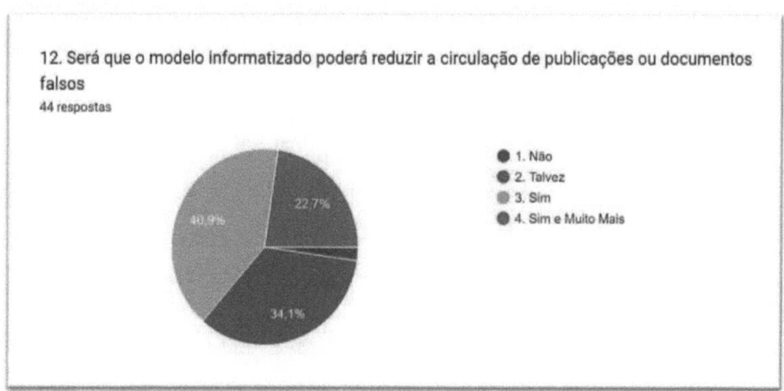

Figure 32: Summary of the fourth question for citizens (Source: *Google Forms*).

In the last question on the impact of the proposed model, both individually and for society, around 84.1% of citizens answered that the model would lead to greater credibility of the information shared, including trust in the information shared in groups or social networks, thus avoiding situations of *fake news*. This coincidence in the answers to questions 13 and 14 clearly shows that once the model has been implemented, citizens and society in general will benefit,

will pay attention to the information shared, since they will know that they can validate the information on a credible platform. This argument is supported by the suggestions made by citizens in question 15, in which the majority praised the initiative and recommended that the model be implemented and disseminated.

In short, it can be seen that for both the citizens' and companies' forms, the respondents are in line with the assumptions of the author's five (5) analysis questions, and also raise the possibility that one of the solutions (the one proposed in this paper) to reduce the level of circulation of supposedly false information necessarily involves the operationalization and dissemination of the model.

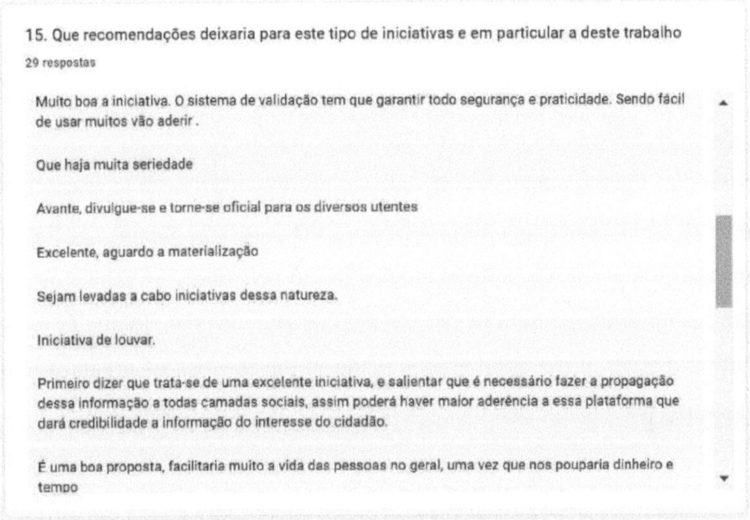

Figure 33: Citizens' recommendations regarding the proposed model (Source: *Google Forms).*

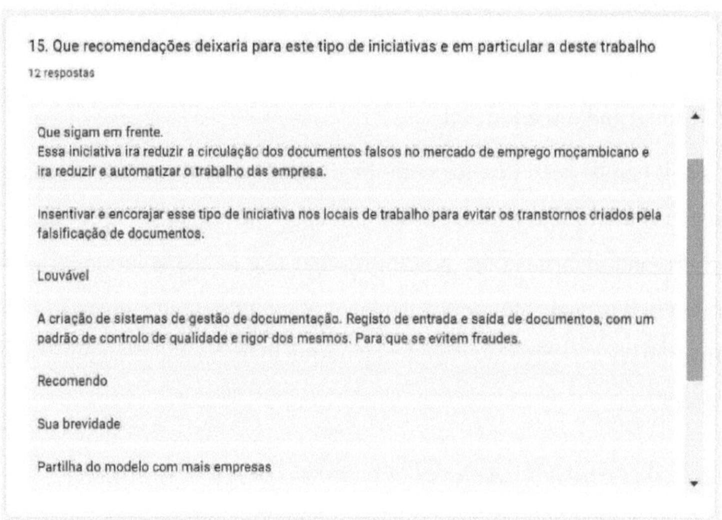

Figure 34: Companies' recommendations regarding the proposed model (Source: *Google Forms*).

Looking at the recommendations left mainly by citizens, there is a certain tendency to consider that the success of the proposed model necessarily depends on how well it is disseminated, the involvement of the agents, the relationship to the context of the institutions and also the simplicity of the solution, based on the assumptions made in the theories of organizational structuring and change and the general theory of information systems.

Thus, by relating the theories and the results of the research, we reach a point of equilibrium, accepting that the sharing of supposedly false information is indeed a problem, but that in order to overcome this problem, we can not only look at the prospect of developing a technological solution as the answer, but we can also look at the component of making the solution viable for acceptance by society. In this approach, the theories of organizational structure and change lead us to think about framing the solution in terms of context and structure. The context in this case makes it possible to find a solution that fits in with the society in which the solution is placed and the structure seen as the hierarchy in the institutions.

In the recommendations put forward by citizens, the fact that the context must once again be taken into *account* when implementing the solution is emphasized: *"the computerized model must be implemented taking into account the reality of the institutions without hindering the execution of processes because not all information is available to verify its reliability"*

With regard to structure, the authors Lessa, Negash and Belachew (2016) cite Bhatnagar (2000) who states that there is a lack of commitment on the part of government and public managers, i.e. the success of the proposed solution necessarily requires the involvement of managers (structure) who can contribute to the correct implementation of the procedures for using the solution. Relating the answers given by the companies to this theory, one of the companies advises coordinated work: *"Coordinated action is necessary and unavoidable"*. In the author's view, the success of the solution will depend on the action taken by managers to disseminate its use in favor of the institution. This strongly supports Bhatnagar's (2000) assertion.

Still on the recommendations left by citizens, some point to the success of the project as being synonymous with the simplicity of the solution: *"The initiative is very good. The validation system must guarantee security and practicality.* This thought can be related to the general theory of information systems in the view of De Araujo *(1995),* when he states that robust solutions can lead to partial or total abandonment, thus requiring solutions to be broken down into small components, i.e. developing simple and practical solutions.

Therefore, spreading the word about the solution (which was emphasized in the comments from both citizens and companies) will allow for greater adherence and acceptance.

Developing a context-oriented solution (emphasized in the comments on the citizens' questionnaire) will create greater involvement and acceptance on the part of the institutions in a specific way.

And developing a simple solution (in terms of steps from entering to obtaining information) could increase adherence, based heavily on the comments in the citizens' questionnaire.

CHAPTER 6

6. CONCLUSIONS AND RECOMMENDATIONS

The conclusions and recommendations chapter provides answers based on the results obtained, all the assumptions made in the objectives, and challenges other researchers to think about the possibility of increasing the gains of this work, by including functionalities that were not explored in the proposed model or that were not a highlight of it.

6.1. Conclusion

The main objective of the work was to propose a computerized model as one of the ways of validating documents and entities, so the model was also developed, mapping local and remote databases, to prove the connectivity between different databases and then validate the information contained in these databases.

In relation to the specific notes of the work, with regard to the traditional ways in which citizens validate documents and entities, the questionnaires showed that around 82% resort to validation with the institutions that supposedly own the advertisements, consultations with colleagues and friends and direct information on the institution's official website. It can be seen, however, that the time taken to validate advertisements and/or information varies mostly between 24 hours and at least 10 days, which is a total loss for those looking for a job vacancy or to validate the agent who comes to their home with the intention of doing a certain job.

As far as organizations (companies) are concerned, just as was the case with individuals, they also resort to validating documents with the entities that produced them and/or by observing the stamp, signature and letterhead. However, there is a certain weakness in the validation mechanisms because, given the sophistication of document tampering platforms, observation with the naked eye can lead to the document being misjudged in terms of its legitimacy.

The proposed model, mapped according to the specific needs of each organization, can take a maximum of 30 seconds to validate a document or entity, and at no additional cost if it doesn't require Internet access. It is clear that the proposed model can help society validate any type of information in real time, as long as the databases are properly mapped.

The problem associated with wasting time and resources (money) is completely solved, as individuals or organizations will be able to validate any information in their possession in 30 seconds or less.

In short, operationalizing the model proposed by the author can help reduce the circulation of supposedly false information, which accounts for 63.6% of the citizen form and around 64.7% of the company form.

6.2.   Recommendations

Today, the big challenge is the availability of real information, which has led many organizations to adopt solutions that provide the public with credible information in real time and at no cost.

The solution proposed in this paper brings with it the challenge of making information available in real time, as long as its data is mapped properly. In the case of mapping different databases from different physical locations and the sensitive information to be shared in the mapping, however, another cybersecurity challenge arises, for a specific *man-in-the-middle* problem, i.e. when passing information from point A to B, within the channel, how can we prevent the information being mirrored in traffic?

67

In an attempt to solve this problem, he recommends:

1. That the solution is administered by a reliable entity (association representing similar institutions);

2. That access to the databases be indicated by a specific IP address (pre-configured) in order to reduce the levels of intrusion into the databases;

3. That the users created and mapped in the application are read-only and;

4. Creating views for the queries required in the model is one of the right measures to avoid direct access to the database tables.

7. Bibliographical references

Almeida, A. P. (2019). The Role of Interoperability in Public Administration: Contributions to improve information management and citizen satisfaction. *Dissertation,* p. 29.

AR, A. d. (July 1, 2012). Draft Revision of the Penal Code. *Book One - General Provisions,* p. 11.

AT. (October 23, 2015). *Mozambique Tax Authority.* Obtained from NUIT Letter Printout: https://nuit.at.gov.mz/nuit/bootstrap/theme/work/Impressao_Carta.as
px

Borges, A. (December 18, 2020). *News.* Retrieved June 1, 2022, from O PAÍS: https://www.opais.co.mz/detidos-supostos-falsificadores-de-documents-in-matola/

LETTER (April 25, 2022). *Economy and Business.* Retrieved June 1, 2022, from Carta de Moçambique: https://www.cartamz.com/~cartamzc/index.php/economia-e-business/item/10486-identified-over-60-fake-engineers-in-mocambique

Chilundo, B. (2020). Institutional Theory & Organizational Change. *Information Systems and Organizational Theories.*

CIN-UFPE (April 23, 2007). *References.* Retrieved July 27, 2022, from Centro de Informática da UFPE: https://www.cin.ufpe.br/~if696/referencias/integracao/_Interoperabil idade_d and_Database_and_Heterogeneous_Systems.pdf

(2013). *Entity Concept.* Retrieved June 30, 2022, from Concept From: https://conceito.de/entidade

DA CUNHA, M. X., JÚNIOR, M. F., MAIA, C. D., & LUCIAN, R. (August 24, 2009). Análise da Implantação dos Sistemas de Informação em uma Instituição Federal de Ensino de Alagoas à Luz da Teoria Institucional . pp. 3 - 4.

DASP (2020). *BAU Information Portal.* Retrieved from e-BAU: http://dasp.mic.gov.mz/contra-prova

De Araujo, V. M. (1995). Information systems: a new theoretical and conceptual approach. p. 31.

DW (2017). *Made for Minds.* Retrieved June 5, 2021, from https://www.dw.com/pt- 002/tribunal-mo%C3%A7ambican-condemns-r%C3%A9us-accused-of-money embezzlement-p%C3%BAblico/a-55340562

Educação, C. E. (2014). *All Articles.* Retrieved June 21, 2022, from Courses School Education: https://cursos.escolaeducacao.com.br/artigo/tipos-de-documents

Efomento (2022). Retrieved June 28, 2022, from Efomento: http : //efomento .cnpq.br/efomento/html/falsiDocPublico.htm

GERENCIAR, G. (January 3, 2020). *Documents.* Retrieved June 23, 2022, from Managing Document Intelligence: https://grupogerenciar.com.br/2020/01/03/como-classificar-os-tipos-de- documents/

I. Assembly of the Republic. (1997). *Lei de Bases das Autarquias* (Law no. 2/97, of May 18th ed.). Maputo: Assembly of the Republic.

Printed (December 24, 2019). Bulletin of the Republic. *Law no. 24/2019: Revision of the penal code,* p. 29.

INE. (2017). Second Census of Mozambican Companies. *Companies in Mozambique: Results of the second national census (2014 - 2015),* pp. 1-3.

INE. (April 29, 2019). IV GENERAL POPULATION AND HOUSING CENSUS - 2017. *GENERAL POPULATION AND HOUSING CENSUS - DEFINITIVE RESULTS,* pp. 39-45.

Infopedia (2022). *Dictionaries.* Retrieved June 21, 2022, from Infopédia-Dicionários Porto Editora: https://www.infopedia.pt/dicionarios/lingua- portuguese/document

INSS (2012). *SISSMO.* Retrieved June 2, 2022, from Instituto Nacional de Segurança Social: https://www.inss.gov.mz/o-inss/o-sissmo.html

Lessa, L., Negash, S., & Belachew, M. (January 2016). Steering E-Government Projects from Failure to Success: Using Design-Reality. p. 5.

Mozambique, P. d. (August 12, 2017). *News.* Retrieved June 1, 2022, from Portal do Governo de Moçambique: https://www.portaldogoverno.gov.mz/por/Imprensa/Noticias/Detecta dos- 177-casos-de-falsificacao-de-documentos

Muquingue, H. (2020). The theory of structuration. *Eduardo Mondlane University.*

Oliveira, K. V. (2014). Archives and Document Security. *Aula,* pp. 10-18.

PortalDoCidadao (2019). *Home.* Retrieved June 2, 2022, from Portal do Cidadão: https: //portalcidadao. gov.mz/

Ravelli, C. A. (2003). Analysis of data interoperability for the implementation of a virtual manufacturing environment. *Dissertation,* p. 30.

SAGE. (2022). *Dictionary definition.* Retrieved June 30, 2022, from
SAGE: https://www.sage.com/pt-pt/blog/dicion%C3%A1rio-termos-
business/entities/#gate-56228f27-e58f-435b-9b72-a561d684e263

MEANINGS (2022). *GENERAL.* Retrieved June 30, 2022, from
SIGNIFICADOS: https://www.significados.com.br/entidade/

SINFIC. (2006). *Articles - Interoperability of Information Systems in the
Modernization of Public Administration.* Retrieved June 1, 2020,
from http : //www. sinfic. pt/SinficWeb/displayconteudo.
do2?numero=24428

(2011). *INSS.* Retrieved from SISSMO:
http://www3.inss.gov.mz/Tesouraria/TesChecarVeracidade/ChecarV
eracidad e?AspxAutoDetectCookieSupport=1

Staff, W. B., Helling, A., Cabannes, Y., Dávila, J., Robson, P., Palmer, I., .
. . Vásconez, J. (2008). *Municipal Development in Mozambique -
Lessons from the First Decade.* Maputp: World Bank Staff.

UCM (2021). *Catholic University of Mozambique - ONLINE Student
Services.* Course Completion Certificate Validation:
https://estudante.ucm.ac.mz/certificate

UNIDO. (February 2002). *The Role of Municipalities in Enterprise
Development - UNIDO Project: YA/MOZ/01/431/11-52
Mozambique.* Vienna: UNIDO.

VIRTUAL, E. (2022). *Document Definition.* Retrieved June 21, 2022, from
Virtual State: https://www.estadovirtual.com.br/o-que-e-documento/

Weimer, B. (2012). *The Tax Base of Mozambican Municipalities:
Characteristics, Potential and Political Economy.* Maputo: IESE.

Zhou, J. (April 12, 2022). *Best PDF converter.* Retrieved from EseaEus:

https://pdf.easeus.com/pdf-converter-tips/best-pdf-converter.html

APPENDICES

1. Data collection form for companies

Proposta do Modelo Informatizado Para a Legitimação de Documentos e Entidades

No âmbito da realização da Dissertação para o Grau de Mestrado em Informática (Engenharia de Software), pela Faculdade de Ciência da Universidade Eduardo Mondlane, submete-se o presente questionário para recolher as sensibilidades das instituições públicas, privadas e Organizações Não-Governamentais sobre a forma de validação de documentos internos e externos tais como: facturas, recibos, crachás – documentos de identificação, certificados, diplomas e outros que, no entendimento do autor, com o modelo a propor, poderá reduzir o nível de circulação de documentos falsos, à medida em que, tornar-se-á possível validar em tempo real mediante a introdução da referência do documento em causa.

Entendendo que as informações partilhadas neste formulário são sensíveis e confidenciais, declaro garantir o anonimato para todos os dados recolhidos, usar a informação para o consumo exclusivo do trabalho em questão e ainda no período de realização do trabalho.

Neste sentido venho agradecer a disponibilidade da V.Excia no preenchimento de forma franca o formulário abaixo.

TEMPO ESTIMADO 5 MINUTOS.

*Obrigatório

1. 1. Nome da Empresa *

2. 2. Tipo de Empresa

   Marcar apenas uma oval.

   - Pública
   - Privada
   - ONG
   - Outra: _______________

3. 3. Área de Actuação

____
____
____
____
____

4. 4. Período de Actuação no Mercado

Marcar apenas uma oval.

- Menos de 6 meses
- Menos de 2 anos
- Menos de 6 anos
- Mais de 6 anos

5. 5. Cargo ou Função do entrevistado

____

6. 6. A empresa produz algum tipo de documento como: factura, recibo, crachá, certificado, notas de encomenda e entrega, diploma etc?

Marcar apenas uma oval.

- Sim
- Não

7. 7. Além dos documentos mencionados acima, a empresa produz outros tipos de documentos, quais?

Marcar apenas uma oval.

- Não
- Outra: _________________________________

8. 8. A empresa já teve a necessidade de validar documentos de outras instituições?

Marcar apenas uma oval.

○ Sim

○ Não

9. 9. E como são validados os documentos produzidos na sua empresa a pedido ou não de outras entidades?

____

____

____

____

____

10. 10. Considera mesmo um problema a partilha de documentos falsos?

Marcar apenas uma oval.

○ Sim

○ Não

○ Talvez

11. 11. Havendo possibilidade de se usar o modelo informatizado para a validação dos documentos da sua empresa, de "1" a "5" qual seria o seu posicionamento.

Marcar apenas uma oval.

○ 1. Sem hipótese

○ 2. Adoptaria com muito receio

○ 3. Adoptaria o modelo

○ 4. Adoptaria e recomendaria a outras entidades

○ 5. Muito excelente, sem comparação

12. 12. Será que o modelo informatizado poderá reduzir a circulação de documentos falsos

Marcar apenas uma oval.

- 1. Não
- 2. Talvez
- 3. Sim
- 4. Sim e Muito Mais

13. 13. Que impacto teria o modelo na sua empresa?

Marcar apenas uma oval.

- 1. Sem nenhum impacto
- 2. Talvez tenha impacto, mas que não me ocorre
- 3. Credibilidade da informação partilhada pela empresa
- 4. Colaboradores educados e atentos a acções fraudulentas de natureza documental
- Outra: _______________________________

14. 14. Que impacto teria o modelo na sociedade?

Marcar apenas uma oval.

- 1. Sem nenhum impacto
- 2. Talvez tenha impacto, mas que não me ocorre
- 3. Credibilidade da informação partilhada
- 4. Sociedade educada e mais confiança nos documentos que circulam
- Outra: _______________________________

15. 15. Que recomendações deixaria para este tipo de iniciativas e em particular a deste trabalho

2. Citizen data collection form

Proposta do Modelo Informatizado Para a Legitimação de Documentos e Entidades

No âmbito da realização da Dissertação para o Grau de Mestrado em Informática (Engenharia de Software), pela Faculdade de Ciência da Universidade Eduardo Mondlane, submete-se o presente questionário para recolher as sensibilidades dos cidadãos sobre a forma de validação de documentos que circulam como: anúncios, concursos, crachás – documentos de identificação, certificados, diplomas e outros que, no entendimento do autor, com o modelo a propor, poderá reduzir o nível de circulação de documentos falsos, à medida em que, tornar-se-á possível validar em tempo real mediante a introdução da referência do documento ou da publicação numa plataforma online
Entendendo que as informações partilhadas neste formulário são sensíveis e confidenciais, declaro garantir o anonimato para todos os dados recolhidos, usar a informação para o consumo exclusivo do trabalho em questão e ainda no período de realização do trabalho.
Neste sentido venho agradecer a disponibilidade da V.Excia no preenchimento de forma franca do formulário abaixo.
TEMPO ESTIMADO 5 MINUTOS.

*Obrigatório

1. 1. Idade *

    Marcar apenas uma oval.

    - 1. 18 - 25 anos
    - 2. 26 - 30 anos
    - 3. 31 - 45 anos
    - 4. Mais de 46 anos

2. 2. Gênero

    Marcar apenas uma oval.

    - 1. Masculino
    - 2. Feminino
    - 3. Outro
    - 4. Não Revelar

3. **3. Nível Académico**

Marcar apenas uma oval.

- 1. MÉDIO GERAL
- 2. MÉDIO PROFISSIONAL
- 3. SUPERIOR
- Outra: _______________

4. **4. Conhece alguém que tenha passado por situação de concorrer a uma vaga inexistente ou aderir uma publicação falsa?**

Marcar apenas uma oval.

- Sim
- Não

5. **5. Já esteve na situação da pergunta anterior?**

Marcar apenas uma oval.

- Sim
- Não

6. **6. Que impacto teve para si a situações acima?**

Marcar tudo o que for aplicável.

- [ ] A. Desperdício do tempo
- [ ] B. Afectou-me moralmente por ter partilhado uma vaga falsa
- [ ] C. Desperdício de recursos (Dinheiro)
- [ ] D. Mal visto nos meus grupos de estudo por ter partilhado informação falsa
- [ ] E. Mal visto no ambiente de Negócio
- [ ] F. O impacto foi incalculável
- [ ] Outra: _______________

7. 7. Nas situações acima, qual foi o método usado para descobrir que a informação era falsa

Marcar apenas uma oval.

- A. Perguntando colegas e amigos
- B. Lendo a informação da página oficial da instituição
- C. Uma publicação da instituição sobre a falsidade do concurso
- D. Visitando as instalações da instituição
- E. Validando numa plataforma credível
- Outra: _______________________________

8. 8. Na situação de publicação de informação falsa, quanto tempo em média leva para validar a informação?

Marcar apenas uma oval.

- A. Menos de 10 minutos
- B. Menos de 5 horas
- C. Dentro de 24 horas
- D. Dentro de 3 dias
- E. Dentro de uma semana
- F. De 10 dias em diante

9. 9. Como cidadão, que documento ou publicação gostaria de ter a possibilidade de validar (BI, certificado de habilitações literais, carta de condução, uma vaga de emprego, concurso, crachá, convites, outros)?

Marcar apenas uma oval.

- A. Nenhum
- B. Pelo menos dois
- C. Todos
- Outra: _______________________________

10. 10. Numa situação de lhe chegar um agente de alguma empresa e se identificar mediante a apresentação do crachá, que mecanismos usa para validar?

Marcar tudo o que for aplicável.

- [ ] A. Observando o logotipo da empresa
- [ ] B. Observando a indumentaria do agente
- [ ] C. Apenas confiando nas boas intenções do agente
- [ ] D. Mediante a validação do código do agente numa plataforma credível
- [ ] Outra: _______________________________

11. 11. Havendo possibilidade de se usar o modelo informatizado para a validação dos documentos ou publicações de entidades, de "1" a "5", qual seria o seu posicionamento.

Marcar apenas uma oval.

- ( ) 1. Sem hipótese
- ( ) 2. Adoptaria com muito receio
- ( ) 3. Adoptaria o modelo
- ( ) 4. Adoptaria e recomendaria a outras entidades
- ( ) 5. Muito excelente, sem comparação

12. 12. Será que o modelo informatizado poderá reduzir a circulação de publicações ou documentos falsos

Marcar apenas uma oval.

- ( ) 1. Não
- ( ) 2. Talvez
- ( ) 3. Sim
- ( ) 4. Sim e Muito Mais

13. 13. Que impacto teria o modelo para si?

Marcar apenas uma oval.

- 1. Sem nenhum impacto
- 2. Talvez tenha impacto, mas que não me ocorre
- 3. Credibilidade da informação partilhada nos meus grupos e indivíduos atentos a situações de Fake News
- Outra: _______________________

14. 14. Que impacto teria o modelo na sociedade?

Marcar apenas uma oval.

- 1. Sem nenhum impacto
- 2. Talvez tenha impacto, mas que não me ocorre
- 3. Credibilidade da informação partilhada, sociedade educada e mais confiança nos documentos que circulam
- Outra: _______________________

15. 15. Que recomendações deixaria para este tipo de iniciativas e em particular a deste trabalho

I want morebooks!

Buy your books fast and straightforward online - at one of world's fastest growing online book stores! Environmentally sound due to Print-on-Demand technologies.

Buy your books online at
www.morebooks.shop

Kaufen Sie Ihre Bücher schnell und unkompliziert online – auf einer der am schnellsten wachsenden Buchhandelsplattformen weltweit! Dank Print-On-Demand umwelt- und ressourcenschonend produziert.

Bücher schneller online kaufen
www.morebooks.shop

info@omniscriptum.com
www.omniscriptum.com

OMNIScriptum

MIX
Papier aus verantwortungsvollen Quellen
Paper from responsible sources
FSC® C105338
FSC
www.fsc.org

Printed by Books on Demand GmbH, Norderstedt / Germany